Art of Problem Solving Presents:

BEAST ACADEMY

MATH
GUIDE
3D

Aligned to the Common Core State Standards

Erich Owen

Jason Batterson

Published by: AoPS Incorporated
                   P.O. Box 2185
                   Alpine, CA 91903-2185
                   (619) 659-1612
                   info@BeastAcademy.com

ISBN: 978-1-934124-46-8

Beast Academy is a registered trademark of AoPS Incorporated.

Written by Jason Batterson
Illustrated by Erich Owen
Colored by Greta Selman
Cover Design by Lisa T. Phan

Visit the Beast Academy website at www.BeastAcademy.com.
Visit the Art of Problem Solving website at www.artofproblemsolving.com.
Printed in the United States of America.
First Printing 2013.

Become a Math Beast!
For additional books,
printables, and more, visit
www.BeastAcademy.com

This is Guide 3D in a four-book series for third grade:

Guide 3A
Chapter 1: Shapes
Chapter 2: Skip-Counting
Chapter 3: Perimeter and Area

Guide 3B
Chapter 4: Multiplication
Chapter 5: Perfect Squares
Chapter 6: The Distributive Property

Guide 3C
Chapter 7: Variables
Chapter 8: Division
Chapter 9: Measurement

Guide 3D
Chapter 10: Fractions
Chapter 11: Estimation
Chapter 12: Area

Contents:

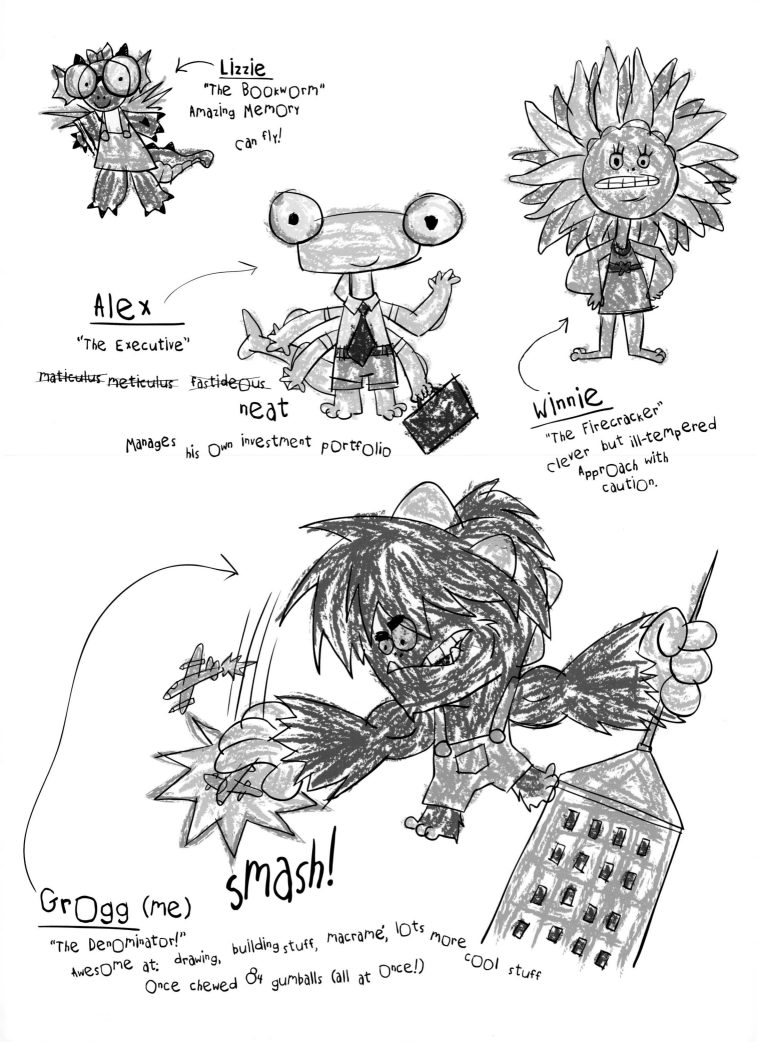

Lizzie
"The BOOKWORM"
Amazing Memory
Can fly!

Alex
"The Executive"
~~maticulus~~ meticulus ~~fastideous~~ neat
Manages his own investment portfolio

Winnie
"The Firecracker"
clever but ill-tempered
Approach with caution.

smash!

GrOgg (me)
"The Denominator!"
Awesome at: drawing, building stuff, macramé, lots more cool stuff
Once chewed 84 gumballs (all at once!)

Welcome to Beast Academy!

This book is called the Guide.

There is also a separate Practice book with lots of problems you can use to sharpen your skills.

The Guide is written like a comic book.

In a comic book, whatever I say shows up in these bubbles. They're called comic balloons.

Here's one!

Each character has a different balloon color. This makes it easy to tell who is talking.

My balloons are purple!

The story is told in panels.

Panels usually have a rectangular frame around them...

...like this one.

Did you notice this stop sign at the bottom of the previous page?

Always try to answer the questions in the stop signs before reading on.

What's this gray box for?

There were 5 panels on page 9.

How many panels are on this page?

STOP SIGNS ASK QUESTIONS THAT YOU SHOULD TRY TO ANSWER BEFORE READING ANY FURTHER.

The gray boxes in the Guide provide important information and definitions.*

Sometimes the gray boxes offer problem solving tips or other useful information.

*AN ASTERISK MAY BE USED TO TIE THE TEXT IN A COMIC BALLOON TO A GRAY BOX.

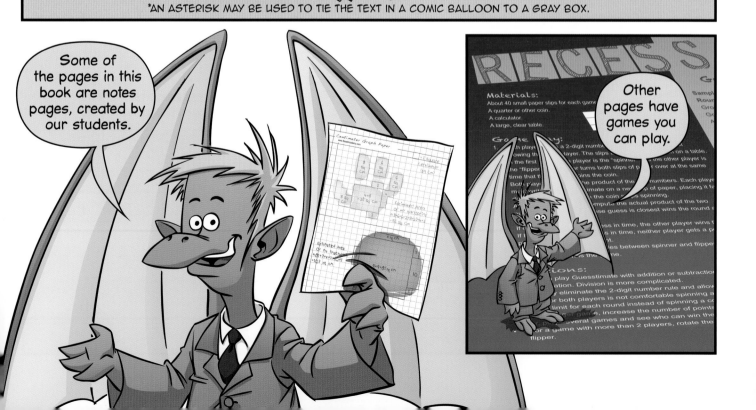

Some of the pages in this book are notes pages, created by our students.

Other pages have games you can play.

Contents: Chapter 10

See page 6 in the Practice book for a recommended reading/practice sequence for Chapter 10.

Chapter 10: Fractions

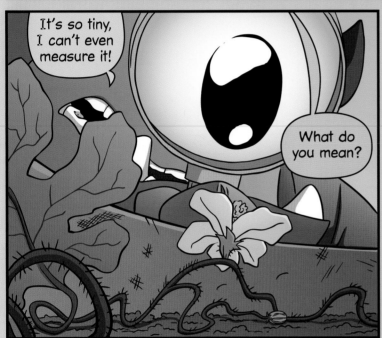

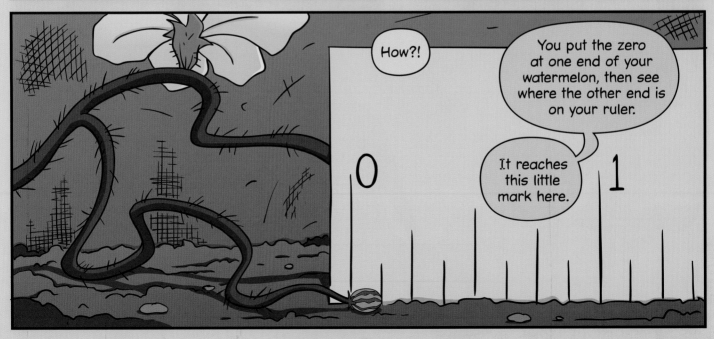

But there are no **numbers** between zero and one on my ruler.

They're not written on your ruler, but there **are** numbers between zero and one.

???

Your watermelon is longer than zero inches, but shorter than one inch.

It's only **part** of an inch long.

Right, so you need a number to describe **what** part of an inch.

0   1

Hmmm...

These little lines on my ruler divide each inch into eight equal pieces.

Since there are eight of them, the pieces are called "eighths."

My watermelon is as long as one piece!

That's right! It's one-eighth inch long.

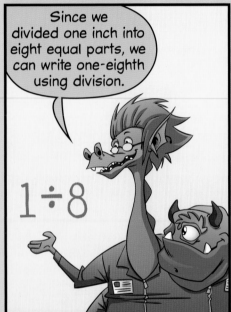

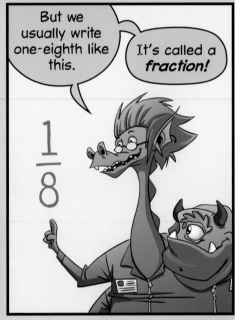

16

To be continued...

What is a fraction?

It's a number.

It's a division.

It's both!

Right! A fraction is a number that is the result of division.

How would we write 1÷3 as a fraction?

The 1 goes on top. It's the number being divided.

The 3 goes on the bottom. It's the number you are dividing by.

It's called the numerator!

It's called the denominator!

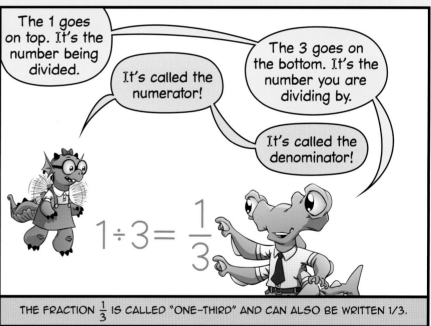

$$1 \div 3 = \frac{1}{3}$$

THE FRACTION $\frac{1}{3}$ IS CALLED "ONE-THIRD" AND CAN ALSO BE WRITTEN 1/3.

Great!

Since a fraction is a number, we can put it on the number line.

Where does $\frac{1}{3}$ belong on the number line?

Where is $\frac{1}{3}$ on the number line?

Let's start with an easier problem.

We can put $\frac{6}{3}$ on the number line.

$\frac{6}{3}$ is $6 \div 3$.

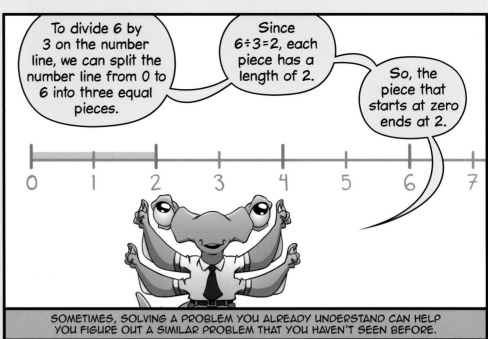

To divide 6 by 3 on the number line, we can split the number line from 0 to 6 into three equal pieces.

Since $6 \div 3 = 2$, each piece has a length of 2.

So, the piece that starts at zero ends at 2.

0   1   2   3   4   5   6   7

SOMETIMES, SOLVING A PROBLEM YOU ALREADY UNDERSTAND CAN HELP YOU FIGURE OUT A SIMILAR PROBLEM THAT YOU HAVEN'T SEEN BEFORE.

To put $\frac{1}{3}$ on the number line, we need to divide 1 by 3.

To divide 1 by 3 on the number line, we can split the number line from 0 to 1 into three equal pieces.

Since $1 \div 3 = \frac{1}{3}$, each piece has a length of $\frac{1}{3}$.

0   $\frac{1}{3}$ 1

The piece that starts at zero must end at $\frac{1}{3}$!

Wonderful! Let's try another.

Where is $\frac{1}{5}$ on the number line?

THE FRACTION $\frac{1}{5}$ IS READ "ONE-FIFTH." $\frac{1}{6}$ IS READ "ONE-SIXTH," $\frac{1}{7}$ IS "ONE-SEVENTH," AND SO ON.

Where is $\frac{1}{5}$ on the number line?

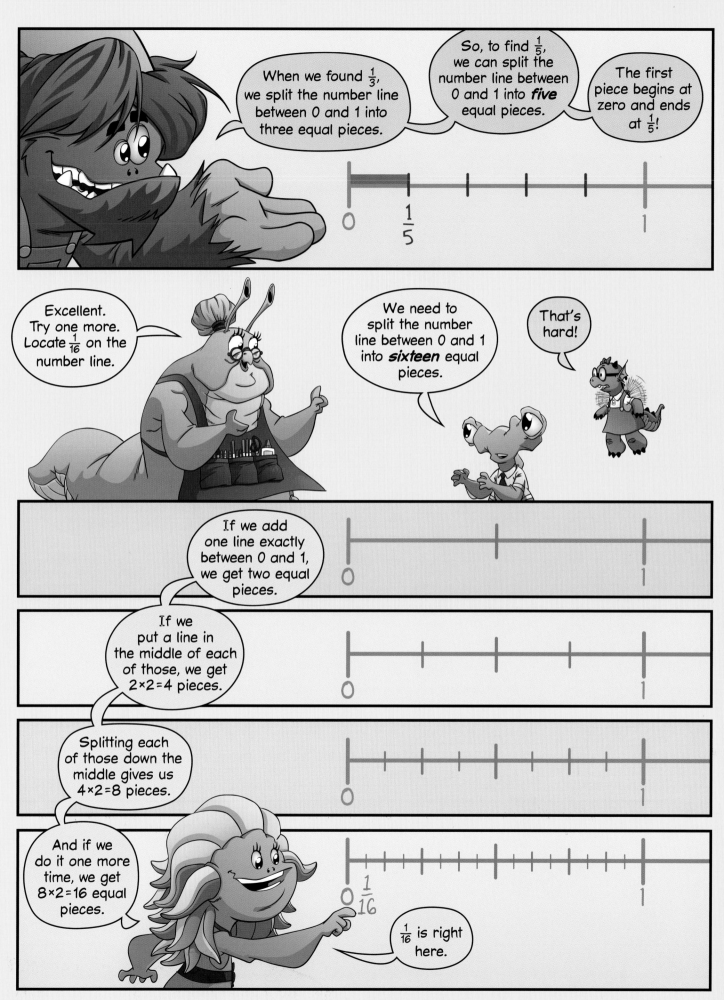

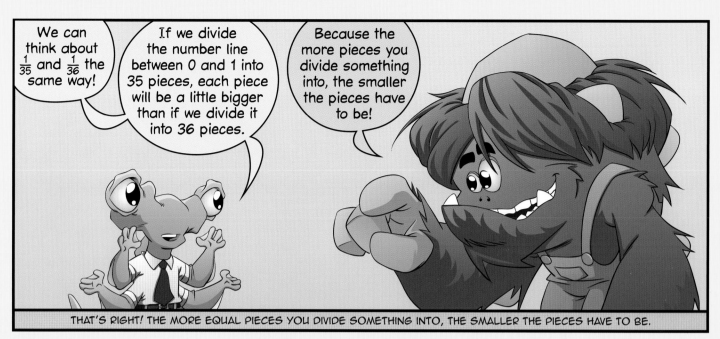

We learned about a new set of numbers today!

They're called fractions!

They're numbers between zero and one.

Great!

But not every fraction is between zero and one.

Some fractions are bigger than 1.

And some fractions are even **equal** to 1!

You just blew my mind.

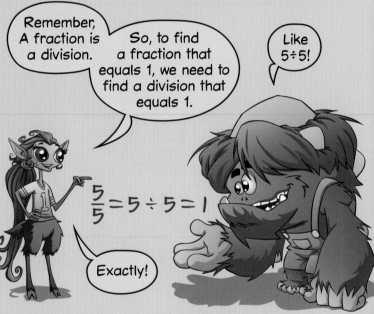

Remember, A fraction is a division.

So, to find a fraction that equals 1, we need to find a division that equals 1.

Like 5÷5!

$$\frac{5}{5} = 5 \div 5 = 1$$

Exactly!

So, any fraction with the same numerator and denominator equals 1.*

Because when you divide a number by itself, the answer is always one.

Great work. All these fractions are equal to 1.

$$\frac{2}{2} = \frac{3}{3} = \frac{4}{4} = \frac{5}{5} = 1$$

Who can think of a fraction that is equal to 2?

Find one.

*EXCEPT $\frac{0}{0}$, WHICH DOESN'T MAKE ANY SENSE BECAUSE WE CAN'T DIVIDE BY ZERO.

For a fraction to equal 3, the numerator has to be three times the denominator.

That's right!

Since $9 \times 3 = 27$, $\frac{27}{9} = 3$.

$$\times 3 \left\{ \frac{n}{9} \qquad n = 27 \right.$$

There are lots of fractions that equal 3.

We can list fractions with whole numbers on the bottom by putting the multiples of 3 on top.

$$\times 3 \left\{ \frac{3}{1} = \frac{6}{2} = \frac{9}{3} = \frac{12}{4} = \frac{15}{5} = \frac{18}{6} = \frac{21}{7} = \frac{24}{8} = \frac{27}{9} \right.$$

We could go on and on forever.

The numbers on top and bottom keep getting bigger, but the fraction still equals 3!

We're counting by 1's on the bottom and skip-counting by 3's on top.

And all the fractions equal 3!

$$= \frac{24}{8} = \frac{27}{9} = \frac{30}{10} = \frac{33}{11} = \frac{36}{12}$$

Exactly!

And if we write fractions that equal five?

The numbers on top and bottom get bigger, but the fractions stay the same.

$$\frac{5}{1} = \frac{10}{2} = \frac{15}{3} = \frac{20}{4} = \frac{25}{5} = \frac{30}{6} =$$

Just because the numbers on the top and bottom of a fraction are big doesn't mean that the fraction is big.

$\frac{999}{999}$ only *seems* like a big number.

But it equals 1!

Practice: Pages 7-11

R&G Equal Fractions

Here we are!

Beast Island's largest smallest fruit and vegetable competition.

Uh-oh!

What's wrong?

I think your watermelon grew.

Maybe it only *looks* bigger.

Let's measure it.

It was one-eighth inch long when we left...

...and now it's one-fourth inch.

Looks like two-eighths to me.

0   1

That's right, but most of the time we call two-eighths one-fourth.

???

Two-eighths and one-fourth are equal!

ONE-FOURTH IS WRITTEN $\frac{1}{4}$, AND IS SOMETIMES CALLED ONE-QUARTER.

Can you explain why $\frac{2}{8}$ and $\frac{1}{4}$ are equal?

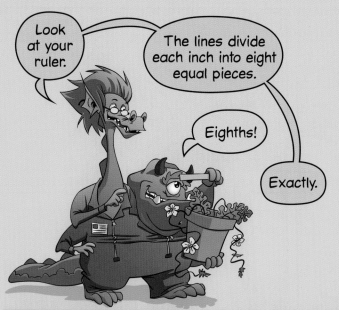

Look at your ruler.

The lines divide each inch into eight equal pieces.

Eighths!

Exactly.

My watermelon is as long as two of these pieces, so it is **two-eighths** of an inch long.

0   $\frac{1}{8}$ $\frac{2}{8}$ 1

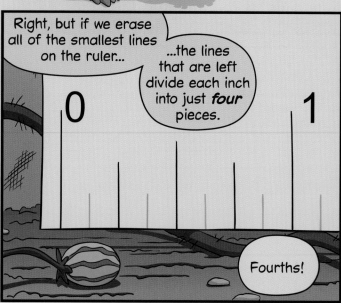

Right, but if we erase all of the smallest lines on the ruler...

...the lines that are left divide each inch into just **four** pieces.

0   1

Fourths!

And since my watermelon is as long as one of these pieces, it's **one-fourth** of an inch long!

0   $\frac{1}{4}$ 1

Yep. So, $\frac{2}{8} = \frac{1}{4}$.

I get it! $\frac{2}{8}$ and $\frac{1}{4}$ are different ways to write the same number.

That's right.

Every fraction can be written lots of ways.

Let's look at $\frac{1}{2}$.

Where is one-half on your ruler?

TWO FRACTIONS THAT ARE DIFFERENT WAYS TO WRITE THE SAME NUMBER ARE CALLED *EQUIVALENT*.

Find $\frac{1}{2}$ on the ruler.

26

28

Zero divided by anything is zero.*

So, $\frac{0}{5}=0$.

And these must be $\frac{2}{5}$, $\frac{3}{5}$, and $\frac{4}{5}$!

$\frac{3}{5}$ is right between $\frac{2}{5}$ and $\frac{4}{5}$!

$\frac{3}{5}$ is a little closer to one than it is to zero.

*EXCEPT $\frac{0}{0}$, WHICH DOESN'T MAKE ANY SENSE.

Excellent observations!

Where is $\frac{6}{5}$ on the number line?

Six-fifths is just one fifth past five-fifths!

It's a little more than one.

That's right.

Is there an easy way to tell if a fraction is greater than one?

If a fraction has the same number on top and bottom, it **equals** one.

If the top number is **bigger** than the bottom number, then the fraction is **greater** than 1.

But if the numerator is **smaller** than the denominator, the fraction is **less** than 1.

Very good. Let's try another.

Where is $\frac{13}{5}$ on the number line?

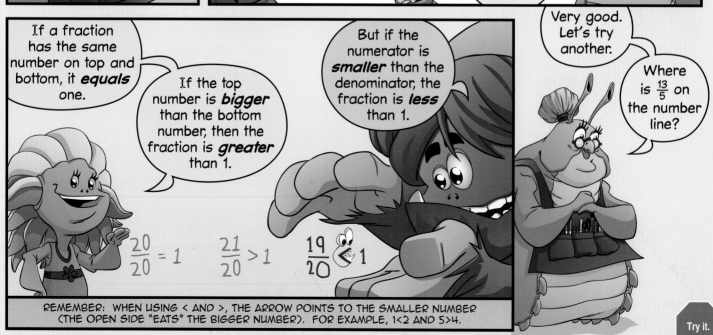

$\frac{20}{20} = 1$ $\frac{21}{20} > 1$ $\frac{19}{20}$ 1

REMEMBER: WHEN USING < AND >, THE ARROW POINTS TO THE SMALLER NUMBER (THE OPEN SIDE "EATS" THE BIGGER NUMBER). FOR EXAMPLE, 1<2 AND 5>4.

Try it.

$\frac{4}{4}$ = 1.

$\frac{8}{4}$ = 2.

$\frac{12}{4}$ = 3.

And $\frac{16}{4}$ = 4.

$\frac{15}{4}$ is between $\frac{12}{4}$ and $\frac{16}{4}$, so it is between 3 and 4.

$\frac{15}{4}$ is three fourths more than 3!

Is $\frac{15}{4}$ equal to $3 + \frac{3}{4}$?

If we start at 3 and count up three more fourths, we get to $\frac{15}{4}$.

So $\frac{15}{4} = 3 + \frac{3}{4}$!

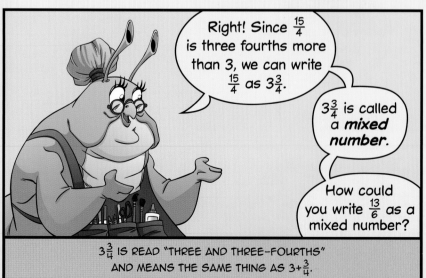

Right! Since $\frac{15}{4}$ is three fourths more than 3, we can write $\frac{15}{4}$ as $3\frac{3}{4}$.

$3\frac{3}{4}$ is called a *mixed number.*

How could you write $\frac{13}{6}$ as a mixed number?

$3\frac{3}{4}$ IS READ "THREE AND THREE-FOURTHS" AND MEANS THE SAME THING AS $3 + \frac{3}{4}$.

$\frac{12}{6}$ equals 2. We need one more sixth to get from $\frac{12}{6}$ to $\frac{13}{6}$.

So $\frac{13}{6} = 2\frac{1}{6}$!

Superb! When a fraction's numerator is greater than its denominator, we can write the fraction as a mixed number.

Why are mixed numbers useful?

SOME PEOPLE CALL A FRACTION WITH A NUMERATOR THAT IS LARGER THAN ITS DENOMINATOR "IMPROPER." SINCE THERE'S NOTHING WRONG WITH "IMPROPER" FRACTIONS, WE PREFER JUST TO CALL THEM FRACTIONS.

It's easier to tell how big the fraction is when it's written as a mixed number.

If a recipe calls for $\frac{7}{3}$ cups of flour, it's easier if it says $2\frac{1}{3}$ cups.

If you are $8\frac{1}{2}$ years old...

...it's confusing if you say that you are $\frac{17}{2}$ years old.

17 quarters equals 4 dollars and one quarter: $\frac{17}{4} = 4\frac{1}{4}$.

And the thirteen half-meatballs in my pocket...

...make $6\frac{1}{2}$ meatballs! Want one?

Grogg!

Practice: Pages 12-21

33

PARTS OF A WHOLE

Fractions be used to mean many different things.

'Tis one o' the reasons fractions be so troublin' to so many monsters.

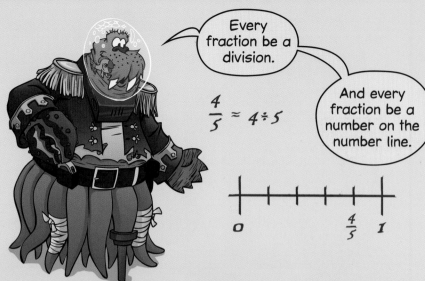

Every fraction be a division.

$$\frac{4}{5} \approx 4 \div 5$$

And every fraction be a number on the number line.

Today we'll learn that a fraction can also be used to mean a part o' somethin'.

Like a half-dollar!

Or a half-pint!

Or a half-nelson!

Aye, all fine examples!

When I split this apple into two equal parts, each part be $\frac{1}{2}$ o' the apple.*

When pirates divide a treasure into equal parts, each part be a fraction of the whole.

REMEMBER, $\frac{1}{2}$ IS A NUMBER. SO, "$\frac{1}{2}$ OF AN APPLE" DESCRIBES AN AMOUNT IN THE SAME WAY THE NUMBER 3 IS USED TO DESCRIBE "3 APPLES."

With nine pirates in me crew, we be dividin' the treasure nine ways.

Each o' the nine parts be $\frac{1}{9}$ of the whole.

'Tis usually a simple task. But some treasures be more difficult to divide than others.

Like what?

Like a priceless paintin'.

How do you divide a *painting* between nine pirates?

By cuttin' it into nine equal rectangles!

Each pirate be gettin' $\frac{1}{9}$ of the whole paintin'.

That's ridiculous!

Aye. 'Twas a poor choice.

Here be my part o' the paintin'...

...$\frac{1}{9}$ of the Mona Beasta.

Other treasures be presentin' different difficulties.

On one voyage, we found two platinum discs.

'Tis no easy task, dividin' two discs equally among nine pirates.

How could the discs be cut so that each pirate be gettin' an equal share?

How could it be done?

If you cut one disc into four pieces...

...and the other disc into five pieces...

...that makes 9 pieces, but they're not all the *same*.

$\frac{1}{4}$ $\frac{1}{4}$ $\frac{1}{4}$ $\frac{1}{4}$

$\frac{1}{5}$ $\frac{1}{5}$ $\frac{1}{5}$ $\frac{1}{5}$ $\frac{1}{5}$

But if you cut both discs into nine pieces, then each pirate could get two pieces.

Each piece is one-ninth of a disc, so each pirate would get two-ninths of a disc.

That makes sense! We divided 2 discs between 9 pirates, and each pirate got two-ninths of a disc!

$2 \div 9 = \frac{2}{9}$!

So, if you divide *three* discs between 9 pirates, each pirate would get $3 \div 9 = \frac{3}{9}$ of a disc!

Aye, $\frac{3}{9}$.
But there be a way to divide three discs without cuttin' each into nine pieces.

How could it be done?

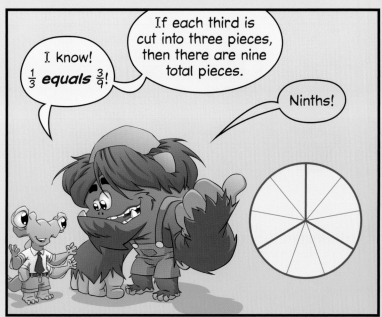

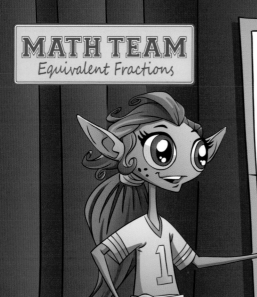

MATH TEAM
Equivalent Fractions

Last practice, we learned to write lots of fractions that equal a whole number, like 5.

Fractions that are equal are called **equivalent** fractions.

$$5 = \frac{5}{1} = \frac{10}{2} = \frac{15}{3} = \frac{20}{4} = \frac{25}{5} = \frac{30}{6} = \dots$$

Like $\frac{1}{3}$ and $\frac{3}{9}$!

$\frac{1}{3}$ and $\frac{3}{9}$ are the same part of a whole.

That's right, Grogg.

Equivalent fractions represent the same part of a whole.

$\frac{1}{3}$

$\frac{3}{9}$

We can also show that two fractions are equivalent using a number line.

Fractions that are the same point on the number line are equivalent.

How could you show that $\frac{1}{3}$ and $\frac{3}{9}$ are the same point on the same number line?

Try it.

If we split each ninth into two equal pieces, we get twice as many pieces!

That makes $2 \times 9 = 18$ pieces all together.

It takes $2 \times 3 = 6$ of these pieces to equal $\frac{3}{9}$.

$$0 \qquad \frac{1}{3} \quad \frac{3}{9} \quad \frac{6}{18} \qquad 1$$

So, we have $\frac{1}{3} = \frac{3}{9} = \frac{6}{18}$!

Making twice as many pieces is the same as multiplying the numerator and the denominator of the fraction by two!

$$\frac{3}{9} \xrightarrow[\times 2]{\times 2} = \frac{6}{18}$$

That's right! Multiplying the numerator and the denominator of a fraction by the same number always creates an equivalent fraction.

$$\frac{5}{8} \xrightarrow[\times 10]{\times 10} =$$

For example we can convert $\frac{5}{8}$ to $\frac{50}{80}$ by multiplying the numerator and denominator by 10.

$$\frac{5}{8} \xrightarrow[\times 10]{\times 10} = \frac{50}{80}$$

Writing a fraction as an equivalent fraction is sometimes called **converting** the fraction.

How could you convert $\frac{6}{10}$ into an equivalent fraction?

$$\frac{6}{10}$$

42

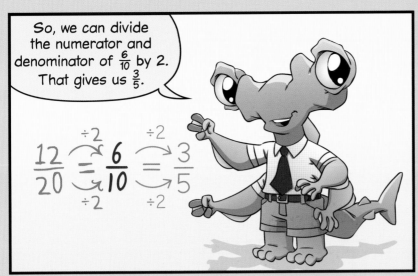

So, we can divide the numerator and denominator of $\frac{6}{10}$ by 2. That gives us $\frac{3}{5}$.

$$\frac{12}{20} \overset{\div 2}{\underset{\div 2}{=}} \frac{6}{10} \overset{\div 2}{\underset{\div 2}{=}} \frac{3}{5}$$

You got it!

When we use division to find an equivalent fraction, it's called *simplifying* the fraction.

Instead of splitting pieces on the number line, we are combining them.

If we put two tenths together, we get one fifth.

So, $\frac{6}{10} = \frac{3}{5}$!

$\frac{6}{10}$

$\frac{3}{5}$

We can only simplify a fraction if the numerator and denominator are both multiples of the same number.

$$\frac{70}{84} \overset{\div 2}{\underset{\div 2}{=}} \frac{35}{42}$$

For example, since 70 and 84 are both multiples of 2, we can simplify $\frac{70}{84}$ by dividing by 2 on top and bottom to get $\frac{35}{42}$.

Then, since, 35 and 42 are both multiples of 7, we can simplify $\frac{35}{42}$ by dividing by 7 on top and bottom.

$$\frac{70}{84} \overset{\div 2}{\underset{\div 2}{=}} \frac{35}{42} \overset{\div 7}{\underset{\div 7}{=}} \frac{5}{6}$$

A fraction is in *simplest form* when 1 is the only whole number that divides both the numerator and denominator without a remainder.

So, in simplest form, $\frac{70}{84}$ is $\frac{5}{6}$.

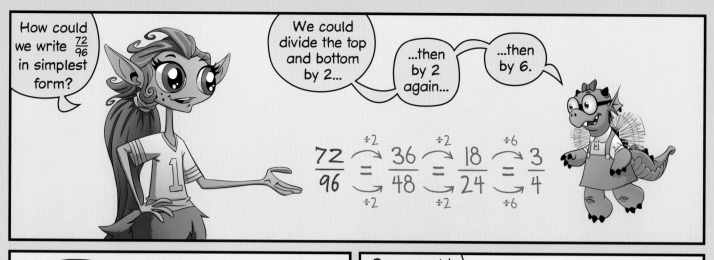

How could we write $\frac{72}{96}$ in simplest form?

We could divide the top and bottom by 2...

...then by 2 again...

...then by 6.

$$\frac{72}{96} \underset{\div 2}{\overset{\div 2}{=}} \frac{36}{48} \underset{\div 2}{\overset{\div 2}{=}} \frac{18}{24} \underset{\div 6}{\overset{\div 6}{=}} \frac{3}{4}$$

Or, we could divide the top and bottom by 2...

...then by 12.

$$\frac{72}{96} \underset{\div 2}{\overset{\div 2}{=}} \frac{36}{48} \underset{\div 12}{\overset{\div 12}{=}} \frac{3}{4}$$

Or, we could divide the top and bottom by 8...

...then by 3.

$$\frac{72}{96} \underset{\div 8}{\overset{\div 8}{=}} \frac{9}{12} \underset{\div 3}{\overset{\div 3}{=}} \frac{3}{4}$$

However we do it, $\frac{72}{96}$ always simplifies to $\frac{3}{4}$!

Simplifying fractions is not so hard.

Now all I need is a way to simplify this thing!

Practice: Pages 22-38

THE LAB

COMPARING FRACTIONS

If two fractions have the same denominator, the one with the larger numerator is bigger.

Five-thirds is bigger than four-thirds...

...because there are more thirds!

$$\frac{5}{3} > \frac{4}{3}$$

But if two fractions have the same numerator, the fraction with the larger denominator is smaller!

Five-eighths is smaller than five-sevenths...

...because eighths are smaller than sevenths.

$$\frac{5}{8} < \frac{5}{7}$$

TURN TO PAGE 21 TO REVIEW WHY EIGHTHS ARE SMALLER THAN SEVENTHS.

But how can we compare fractions whose numerators and denominators are different? For example, which is larger: $\frac{7}{9}$ or $\frac{2}{3}$?

Maybe we could convert one of the fractions so that both fractions have the **same** numerator or denominator.

$$\overset{\times 3}{\underset{\times 3}{\frac{2}{3} = \frac{6}{9}}}$$

We can convert $\frac{2}{3}$ to $\frac{6}{9}$ by multiplying the top and bottom by 3.

Since $\frac{6}{9}$ is less than $\frac{7}{9}$...

...we know that $\frac{2}{3}$ is less than $\frac{7}{9}$.

$$\frac{6}{9} < \frac{7}{9}$$

$$\frac{2}{3} < \frac{7}{9}$$

Quite ingenious!

When comparing two fractions, we can convert one or both fractions.
Try this one: which is larger, $\frac{3}{4}$ or $\frac{17}{20}$?

$$\frac{3}{4} \qquad \frac{17}{20}$$

Since 20 is a multiple of 4, we can multiply the top and bottom of $\frac{3}{4}$ by 5 so that its denominator is 20.

Since $\frac{15}{20}$ is less than $\frac{17}{20}$...

$$\begin{array}{c} \times 5 \\ \frac{3}{4} \rightleftharpoons \frac{15}{20} \\ \times 5 \end{array}$$

$$\frac{15}{20} < \frac{17}{20}$$

$$\frac{3}{4} < \frac{17}{20}$$

...$\frac{3}{4}$ is less than $\frac{17}{20}$.

Well done! We usually convert the fractions so that their denominators are the same.

But sometimes it makes more sense to convert the fractions so that their numerators are equal.

How could we compare $\frac{4}{5}$ to $\frac{8}{11}$?

$$\frac{4}{5} \qquad \frac{8}{11}$$

Try it!

We can convert $\frac{4}{5}$ to $\frac{8}{10}$ by multiplying the top and bottom by 2.

$$\frac{4}{5} = \frac{8}{10}$$
×2 ... ×2

Since $\frac{8}{10}$ is greater than $\frac{8}{11}$...

$$\frac{8}{10} > \frac{8}{11}$$

...$\frac{4}{5}$ is greater than $\frac{8}{11}$!

$$\frac{4}{5} > \frac{8}{11}$$

Right!

Klik!

Oh no...

...not again.

Bwah Ha Ha! Professor Grok is gone. I've abducted your educator. It's time for something much more diabolically difficult!

Furiously Frustrating Fractions!

Comparing **two** fractions is frightfully facile.

Placing three fractions in order is potentially puzzling.

But these **five** fractions create a conundrum with...

Ruinous Ramifications.

$$\frac{15}{16} \quad \frac{16}{15} \quad \frac{29}{32} \quad \frac{31}{29} \quad \frac{33}{31}$$

I've locked your lecturer in the lower level laundry room.

To crack the code, place these five fractions in order from least to greatest.

The combination to your lecturer's lock will be enumerated in the numerators.

While the correct combination will unlatch the lock, the incorrect combination will cause...

Catastrophic Consequences!

Can you put all five fractions in order?

50

Practice: Pages 39-45

Alex

Ones $\frac{0}{1}$ $\frac{1}{1}$

Halves $\frac{1}{2}$

Thirds $\frac{1}{3}$ $\frac{2}{3}$

Fourths $\frac{1}{4}$ $\frac{3}{4}$

Fifths $\frac{1}{5}$ $\frac{2}{5}$ $\frac{3}{5}$ $\frac{4}{5}$

Sixths $\frac{1}{6}$ $\frac{5}{6}$

Sevenths $\frac{1}{7}$ $\frac{2}{7}$ $\frac{3}{7}$ $\frac{4}{7}$ $\frac{5}{7}$ $\frac{6}{7}$

Eighths $\frac{1}{8}$ $\frac{3}{8}$ $\frac{5}{8}$ $\frac{7}{8}$

Ninths $\frac{1}{9}$ $\frac{2}{9}$ $\frac{4}{9}$ $\frac{5}{9}$ $\frac{7}{9}$ $\frac{8}{9}$

Tenths $\frac{1}{10}$ $\frac{3}{10}$ $\frac{7}{10}$ $\frac{9}{10}$

Elevenths $\frac{1}{11}$ $\frac{2}{11}$ $\frac{3}{11}$ $\frac{4}{11}$ $\frac{5}{11}$ $\frac{6}{11}$ $\frac{7}{11}$ $\frac{8}{11}$ $\frac{9}{11}$ $\frac{10}{11}$

Twelfths $\frac{1}{12}$ $\frac{5}{12}$ $\frac{7}{12}$ $\frac{11}{12}$

Contents: Chapter 11

See page 46 in the Practice book for a recommended reading/practice sequence for Chapter 11.

Chapter 11:
Estimation

How many monsters are there on Beast Island?

Right now?

Sure.

Do we count as one monster or two?

???

Monsters come and go all the time. I don't think I could count them all.

You might not know the *exact* number, but you could *estimate.*

Huh?

Estimating means using what you know to make a good guess.

You want me to guess?

Right, but put some thought into it.

TO ESTIMATE MEANS TO FIND A NUMBER THAT IS CLOSE TO EXACT. **AN ESTIMATE** IS A NUMBER THAT IS CLOSE TO AN EXACT AMOUNT.

Okay, then, I figure there are about 503,702 beasts on the island.

How'd you come up with *that* number?

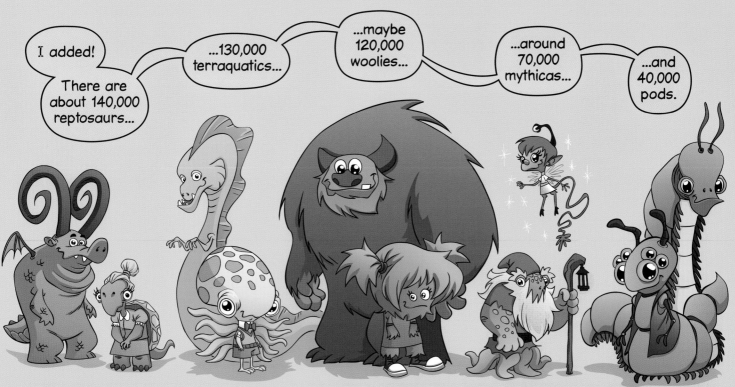

I added! There are about 140,000 reptosaurs...

...130,000 terraquatics...

...maybe 120,000 woolies...

...around 70,000 mythicas...

...and 40,000 pods.

That's 500,000. What about the other 3,702?

```
  140,000
  130,000
  120,000
   70,000
+  40,000
  500,000
```

There are about 3,000 avians...

...700 cybots...

...and us! That makes 503,702 beasts.

That's pretty good, but when you *estimate*, you usually *round* your answer.

???

WE'LL EXPLAIN *HOW* TO ROUND IN THE NEXT SECTION.

A rounded answer usually ends in one or more zeroes.

That makes sense. Zero *is* the roundest number.

55

You already rounded the numbers for each type of beast...

...140,000 reptosaurs, 130,000 terraquatics, and 120,000 woolies.

The numbers were a *lot* easier to add that way!

Right. Making numbers easier to work with is one good reason to round.

But you didn't round your *answer.*

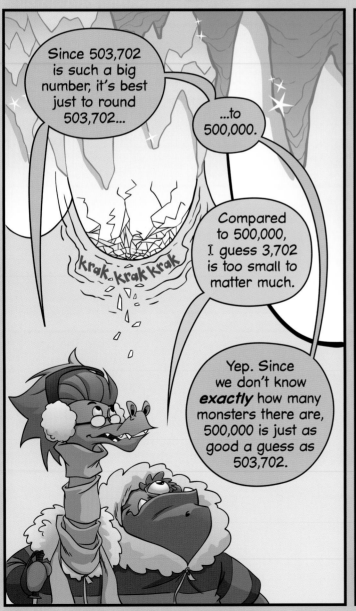

Since 503,702 is such a big number, it's best just to round 503,702...

...to 500,000.

Compared to 500,000, I guess 3,702 is too small to matter much.

Yep. Since we don't know *exactly* how many monsters there are, 500,000 is just as good a guess as 503,702.

krak krak krak

So we can say that there are about 500,000 monsters on Beast Island?

Yep.

What if a new one is born?

One monster is not enough to change our estimate.

Good.

Isn't it cute?

krak krak

Ms. Q. Rounding

When do we estimate?

When we don't need an exact answer.

Or to make a prediction.

Or, when an exact answer is impossible to find.

Like the number of planets in the universe.

Or how many licks it takes to get to the center of a Tootsie Pop.

1...

2...

3...

KRUNCH

The world may never know.

How do we estimate?

Sometimes we estimate by rounding.

We can round to the nearest ten, hundred, thousand, million...

...or any other place value.

That's right. How would you round 781 to the nearest hundred?

Try it.

781 is closer to 800 than to 700.

So, 781 rounds to 800.

Good! We say that 781 rounds *up* to 800, since the rounded number is larger than the original.

How would you round 11,499 to the nearest *thousand?*

11,499 is between 11,000 and 12,000.

Since 11,500 is halfway between 11,000 and 12,000...

...11,499 is closer to 11,000.

So, 11,499 rounds *down* to 11,000!

How would we round 11,500? It's not *closer* to 11,000 *or* 12,000.

It's right in the middle.

The rule is, "Numbers that are exactly in the middle get rounded *up.*"

11,500 rounds up to 12,000.

Why is *that* the rule?

That way, you only need to look at one digit to know which way to round!

What do you mean, Winnie?

To round to the nearest *thousand*, if the hundreds digit is a 5 or higher, we round up...

...and if it's a 4 or lower, we round down.

We only need to look at the *hundreds* digit, because 500 rounds up!

To round to the nearest *hundred*, we only need to look at the *tens* digit.

If the tens digit is a 5 or higher, we round up to the next hundred.

But if the tens digit is a 4 or lower, we round down.

	rounds to
693	700
267	300
950	1000

	rounds to
607	600
221	200
949	900

To round to the nearest *ten*, we look at the *ones* digit.

92 rounds to 90

To round to the nearest *hundred*, we look at the *tens* digit.

437 rounds to 400

And to round to the nearest *thousand*, we look at the *hundreds* digit.

5782 rounds to 6000

Whatever place value we are rounding to...

...the digit to the right tells us whether to round up or down.

THE SYMBOL ≈ MEANS "IS APPROXIMATELY" OR "IS ABOUT."

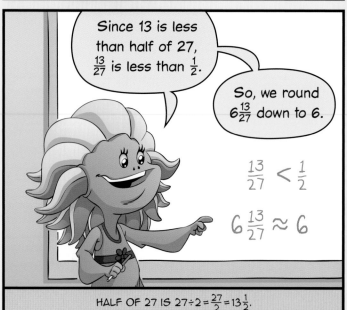

HALF OF 27 IS $27 \div 2 = \frac{27}{2} = 13\frac{1}{2}$.

MATH TEAM
Over and Underestimating

It is often useful to know whether an estimate is bigger or smaller than the actual value.

Let's begin by estimating this sum and this product.

$$4\frac{7}{9} + 1\frac{5}{6} \qquad 613 \times 41$$

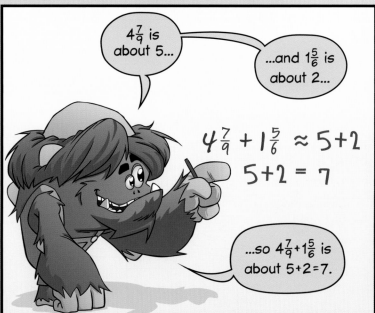

$4\frac{7}{9}$ is about 5...

...and $1\frac{5}{6}$ is about 2...

$$4\frac{7}{9} + 1\frac{5}{6} \approx 5+2$$
$$5+2 = 7$$

...so $4\frac{7}{9} + 1\frac{5}{6}$ is about $5+2 = 7$.

613 is a little more than 600...

...and 41 is just over 40...

$$613 \times 41 \approx 600 \times 40$$
$$600 \times 40 = 24{,}000$$

...so 613×41 is about $600 \times 40 = 24{,}000$.

Good work. Are these estimates larger or smaller than the actual answers?

$$4\frac{7}{9} + 1\frac{5}{6} \approx 5+2$$
$$5+2 = 7$$

$$613 \times 41 \approx 600 \times 40$$
$$600 \times 40 = 24{,}000$$

Which estimate is larger than the actual value? Which is smaller?

63

Grogg rounded both numbers he was adding **up**.

This made his estimate larger than the real answer.

$$4\tfrac{7}{9} + 1\tfrac{5}{6} \approx 5 + 2$$
$$5 + 2 = 7$$

Right!

When an estimate is larger than the actual value, it is called an **over**estimate.

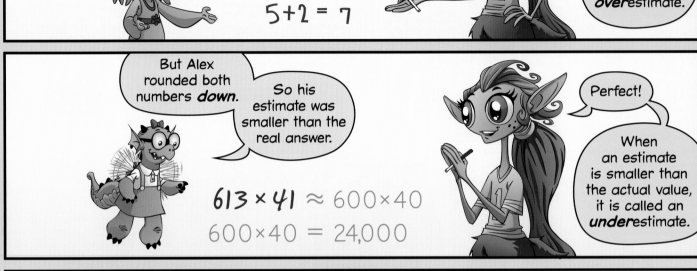

But Alex rounded both numbers **down**.

So his estimate was smaller than the real answer.

$$613 \times 41 \approx 600 \times 40$$
$$600 \times 40 = 24{,}000$$

Perfect!

When an estimate is smaller than the actual value, it is called an **under**estimate.

When you add or multiply two numbers, rounding them up makes the estimate larger than the real answer.

And rounding them both down makes the estimate smaller than the real answer.

Real Answer $\quad 4\tfrac{7}{9} + 1\tfrac{5}{6} = 6\tfrac{11}{18}$

Over-estimate $\quad 5 + 2 = 7$

Real Answer $\quad 613 \times 41 = 25{,}133$

Under-estimate $\quad 600 \times 40 = 24{,}000$

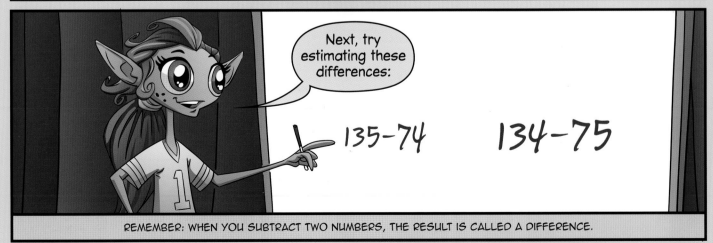

Next, try estimating these differences:

$$135 - 74 \qquad 134 - 75$$

REMEMBER: WHEN YOU SUBTRACT TWO NUMBERS, THE RESULT IS CALLED A DIFFERENCE.

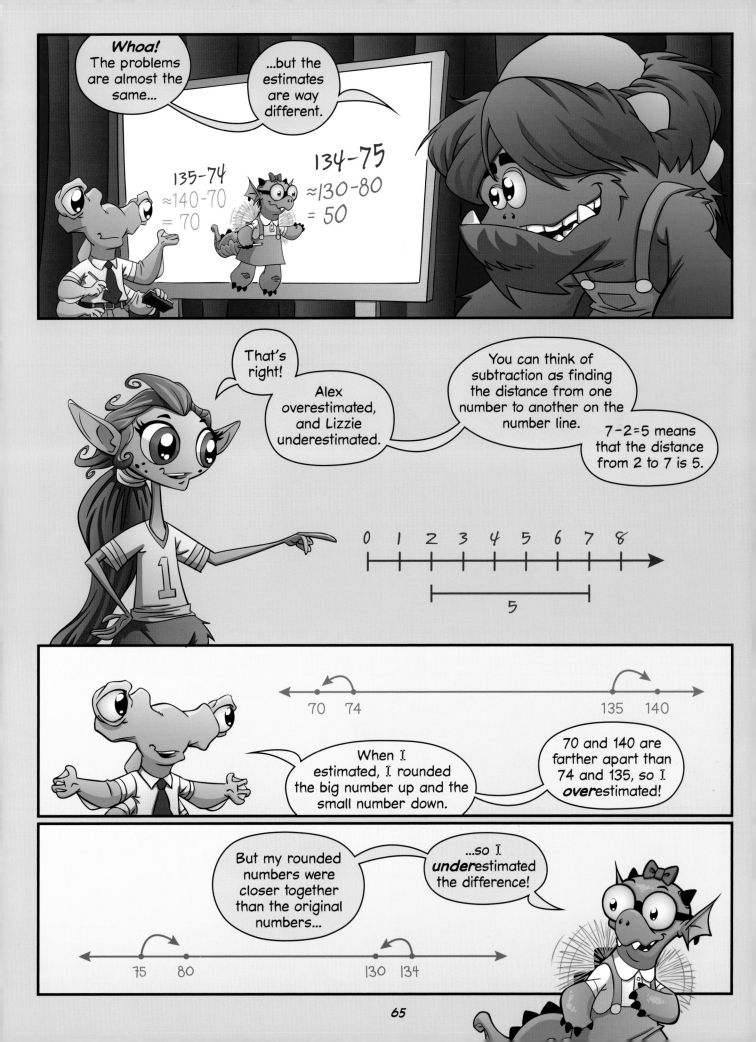

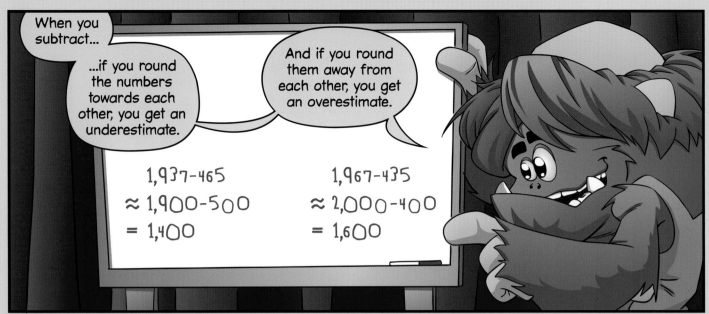

When you subtract...

...if you round the numbers towards each other, you get an underestimate.

And if you round them away from each other, you get an overestimate.

$$1{,}937 - 465$$
$$\approx 1{,}900 - 500$$
$$= 1{,}400$$

$$1{,}967 - 435$$
$$\approx 2{,}000 - 400$$
$$= 1{,}600$$

It can be useful to know whether you are underestimating or overestimating.

That way, you know if the *real* answer is higher or lower than your estimate.

That's right!

Estimate 23×44.

Since 20×40=800 is an *under*estimate...

...23×44 is *more than* 800.

Try 243−67.

Since 240−70=170 is an under-estimate...

...243−67 is *more than* 170.

I didn't know you had it in you, Grogg.

Never underestimate the mind of a great big purple furball.

Practice: Pages 47-61

Once in a blue moon, fortune be smilin' upon me.

Years ago, we discovered a booty so bountiful,

so massive,

so enormous,

so behemoth,

so immense,

so gigantic...

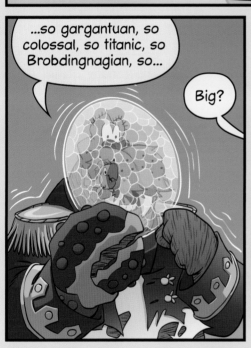

...so gargantuan, so colossal, so titanic, so Brobdingnagian, so...

Big?

Aye, 'twas big.

Too many coins to count.

If you couldn't count the coins, how could you split them into equal shares?

By estimatin'!

It took nearly 100 full wheelbarrows to transport the treasure to the hull o' me ship.

A full 'barrow holds about 180 pounds o' gold coins.

There be about 20 coins in a pound.

Who can estimate the number o' coins in the whole treasure?

Try it.

If 100 wheelbarrows each hold 180 pounds of coins...

...that makes about 100×180=18,000 pounds of coins.

100 wheelbarrows
× 180 pounds in each
= 18,000 pounds

And if each pound was about 20 coins...

...there were about 360,000 coins!

18,000 pounds
× 20 coins in each
= 360,000 coins

REMEMBER: TO MULTIPLY 18,000×20, WE CAN MULTIPLY 18×2, THEN PUT FOUR ZEROS AT THE END.

Aye, about 360,000 coins in all.

After dividin' the coins among nine pirates...

...how many coins be there in each pirate's share?

Since 36÷9 is 4, we can divide 360,000÷9!

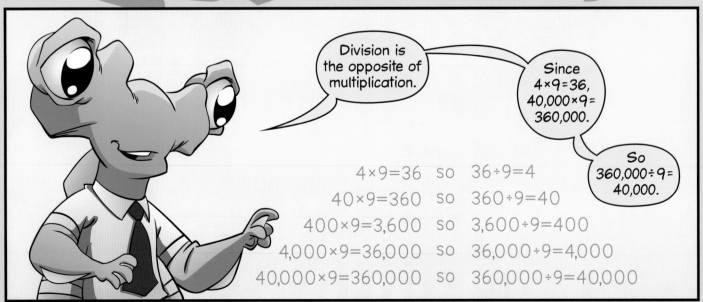

Division is the opposite of multiplication.

Since 4×9=36, 40,000×9= 360,000.

So 360,000÷9= 40,000.

4×9=36   so   36÷9=4
40×9=360   so   360÷9=40
400×9=3,600   so   3,600÷9=400
4,000×9=36,000 so   36,000÷9=4,000
40,000×9=360,000 so 360,000÷9=40,000

We need to estimate 700÷9.

We can round 9 up to 10 and divide 700÷10 to get **70**.

Or, since 700 is close to 720, 720÷9=**80** is also a good estimate of 700÷9.

$$700 \div 9$$
$$\approx 700 \div 10$$
$$= 70$$

$$700 \div 9$$
$$\approx 720 \div 9$$
$$= 80$$

So, each pirate got about 70 or 80 cups of coins.

Aye! Excellent estimatin'!

Of course, no pirate be content with just an estimate, so we had to be countin' all 704 cups o' coins.

After all the cups be divided equally, each pirate be gettin' 78 cups o' coins, with two left over.

How did you decide who got the last two cups of coins?

With a rock-paper-scissors tournament.

Did you win?

Arrr... I lost to Rocky in the third round.

'Twas probably a mistake to play with me right hand.

71

Practice: Pages 62-75

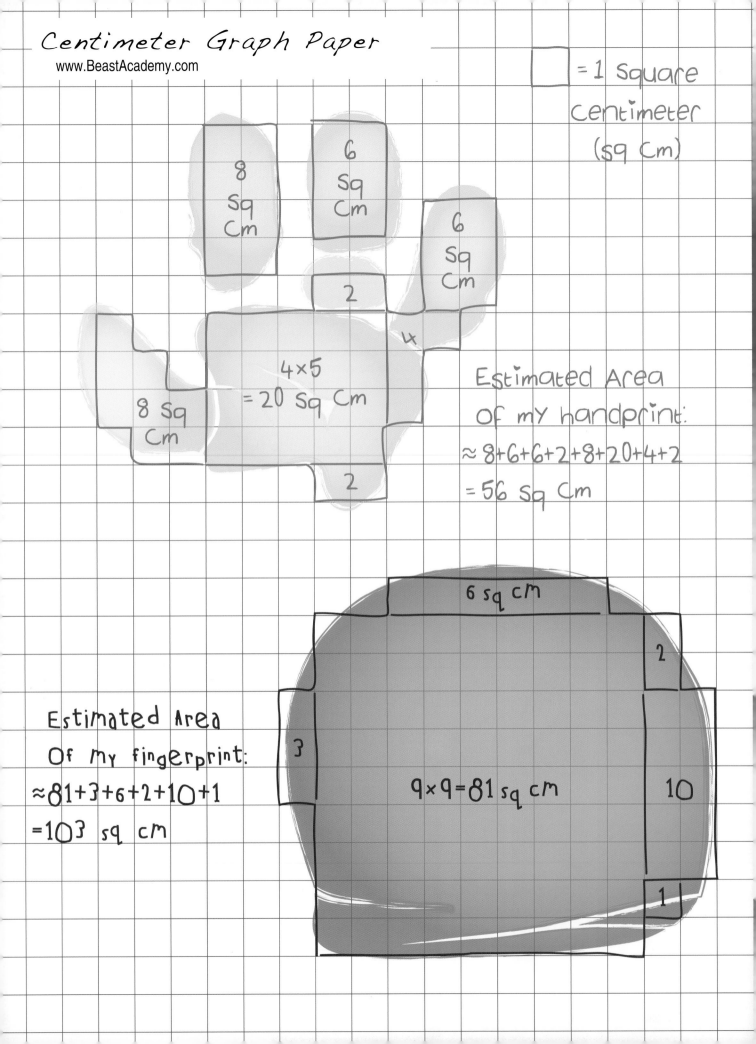

Centimeter Graph Paper
www.BeastAcademy.com

☐ = 1 Square centimeter (sq Cm)

8 Sq Cm

6 Sq Cm

6 Sq Cm

2

4

4×5 = 20 Sq Cm

8 Sq Cm

2

Estimated Area of my handprint:
≈ 8+6+6+2+8+20+4+2
= 56 Sq Cm

6 sq cm

2

3

9×9 = 81 sq cm

10

1

Estimated Area of my fingerprint:
≈ 81+3+6+2+10+1
= 103 sq cm

RECESS

Materials:

About 40 small paper slips for each game.

A quarter or other coin.

A calculator.

A large, clear table.

Game Play:

1. Each player writes a 2-digit number on a slip of paper without showing the other player. The slips are placed face-down on a table.

2. In the first round, one player is the "spinner" and the other player is the "flipper." The flipper turns both slips of paper over at the same time that the spinner spins the coin.

3. Both players estimate the product of the two numbers. Each player must write his or her estimate on a new slip of paper, placing it face down on the table before the coin stops spinning.

4. A calculator is used to compute the actual product of the two numbers. The player whose guess is closest wins the round and gets a point.

5. If one player does not guess in time, the other player wins the round. If both players fail to guess in time, neither player gets a point. If the players tie, both get a point.

Each round, players switch roles between spinner and flipper. The first player to 3 points wins the game.

Variations:

- You can play Guesstimate with addition or subtraction instead of multiplication. Division is more complicated.

- You may eliminate the 2-digit number rule and allow larger numbers.

- If one or both players is not comfortable spinning a coin, you may set a time limit for each round instead of spinning a coin.

- For a longer game, increase the number of points needed to win. Or, play several games and see who can win the most.

- For a game with more than 2 players, rotate the roles of spinner and flipper.

Guesstimate

Sample Game:

Round 1

Grogg writes 28, Alex writes 77.

Grogg estimates 2,400.

Alex estimates 2,300.

Correct product: 28×77=2,156.

Alex gets a point.

Round 2

Grogg writes 61, Alex writes 44.

Grogg estimates 2,600.

Alex estimates 2,800.

Correct product: 61×44=2,684.

Grogg gets a point.

Round 3

Grogg writes 92, Alex writes 15.

Grogg estimates 1,350.

Alex estimates 1,400.

Correct product: 92×15=1,380.

Alex gets his second point.

Round 4

Grogg writes 36, Alex writes 21.

Grogg estimates 800.

Alex estimates 750.

Correct product: 36×21=756.

Alex gets his third point and wins the game.

Find a partner and play!

Contents: Chapter 12

See page 76 in the Practice book for a recommended reading/practice sequence for Chapter 12.

Chapter 12:
Area

Square feet and **square yards** are units of **area!**

Huh?

If we want to know how much space something takes up, we can divide it into squares.

Sure, but what does that have to do with feet?

A **foot** is a unit of length. It's about this long.

A **square foot** is the **area** of a square with sides that are one foot long.

Each of these tiles covers one square foot.

So, the monster in flooring wasn't insulting me?

Nope. He just wanted to know how big our kitchen is.

Let's find out! I'll start counting the tiles.

But...

What is the best way to find the area of the kitchen?

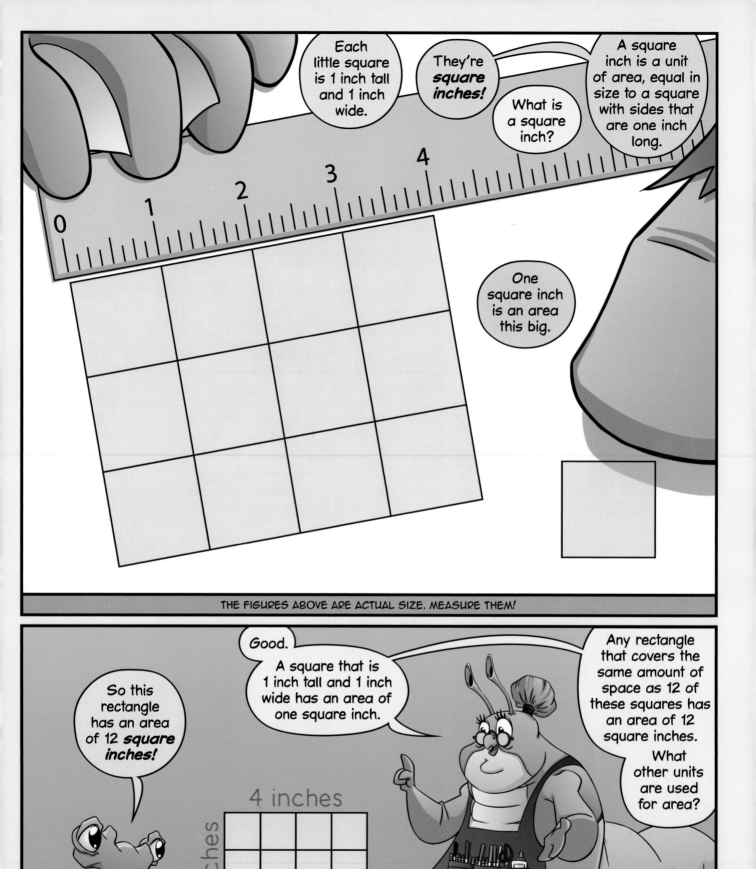

THE FIGURES ABOVE ARE ACTUAL SIZE. MEASURE THEM!

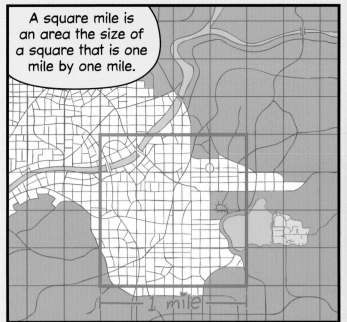

A square mile is an area the size of a square that is one mile by one mile.

1 mile

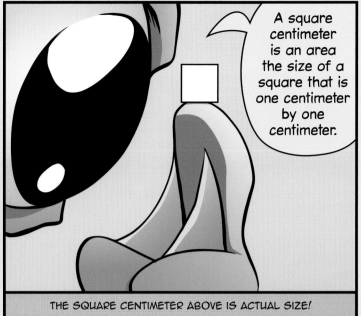

A square centimeter is an area the size of a square that is one centimeter by one centimeter.

THE SQUARE CENTIMETER ABOVE IS ACTUAL SIZE!

Every unit of length has its own unit of area.

There are square inches, square feet, square yards, square miles...

...and metric units like square centimeters, square meters, and square kilometers.

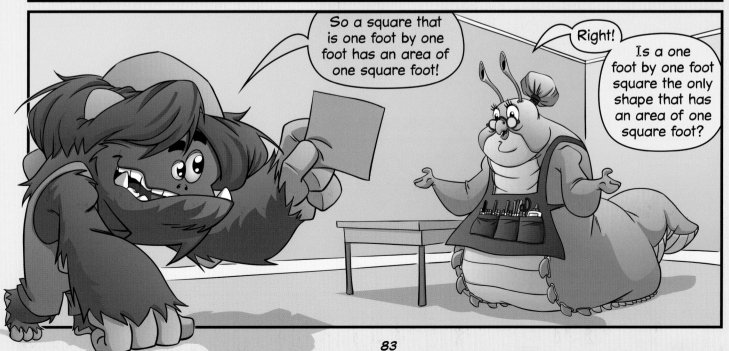

So a square that is one foot by one foot has an area of one square foot!

Right!

Is a one foot by one foot square the only shape that has an area of one square foot?

83

THIS IS AN EXAMPLE OF A *TANGRAM* PUZZLE. A SQUARE IS CUT AS SHOWN INTO SEVEN PIECES THAT CAN BE REARRANGED TO FORM A NEW SHAPE. CUT OUT YOUR OWN SET OF TANGRAM PIECES AT BEASTACADEMY.COM.

Any shape that takes up the same amount of space as a one foot by one foot square has an area of one square foot.

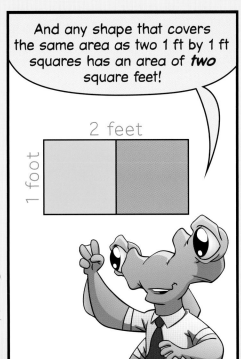

And any shape that covers the same area as two 1 ft by 1 ft squares has an area of *two* square feet!

2 feet

1 foot

Like PolyGrogg!

What!?

It's me, but in polygon form!

I made him from the pieces of two one foot by one foot squares!

So, PolyGrogg has an area of two square feet!

PolyGrogg! What will you little monsters think of next?

Practice: Pages 77-84

Tangrams

Tangrams are an ancient puzzle form in which a square is cut into 7 pieces called tans that can be arranged to make new shapes.

"Among all the kinds of serpents, there is none comparable to the Dragon."

TANGONS

by Lizzie

Tans can be arranged to make a shape that looks like a monster, a building, a letter, or even a dragon. Each of these dragons was made with one or two sets of tans. I made dragons because I am a dragon and I think dragons are awesome.

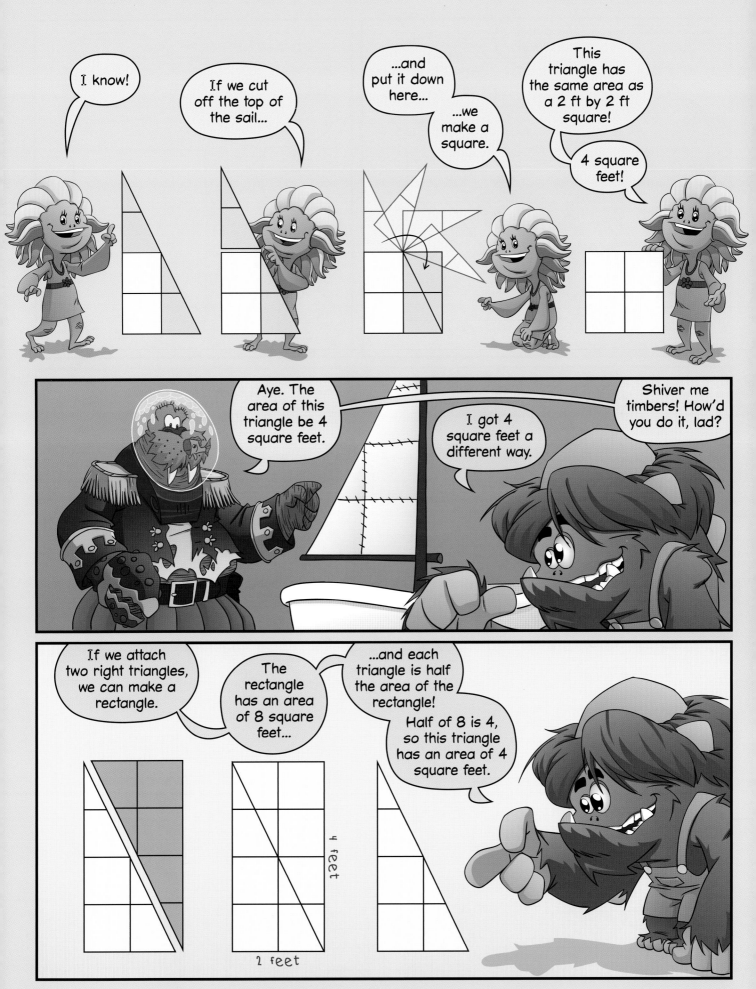

89

Good thinkin'.

Let's try another. What be the area of this right triangle?

6 feet

5 feet

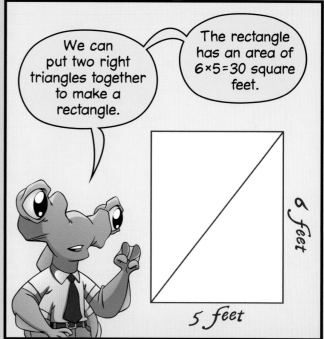

We can put two right triangles together to make a rectangle.

The rectangle has an area of 6×5=30 square feet.

6 feet

5 feet

Since there are two triangles, each triangle has an area of 30÷2=15 square feet.

6 feet

5 feet

Aye. To find the area of a right triangle, 'tis easiest to think of it as half of a rectangle.

So, to find the area of a right triangle...

To find the area of a rectangle, we multiply its width by its height.

...we multiply its width by its height, then divide by 2.

height

width

Area of a rectangle
= width × height

height

width

Area of a right triangle
= width × height ÷ 2

Very good. These be called **formulas.**

A formula is an equation that we can use for a particular type of problem...

...like findin' the area of a right triangle.

We can shorten formulas by usin' variables, like A for area, w for width, and h for height.

Area of a rectangle = width × height

$$A = w \times h$$

Area of a right triangle = width × height ÷ 2

$$A = w \times h \div 2$$

Try this tricky triangle area problem.

Here be a triangle with sides of length 3, 4, and 5 feet.

This be a **right** triangle.*

Try findin' the area of this right triangle.

Try it.

THREE SIDE LENGTHS DO NOT ALWAYS MAKE A RIGHT TRIANGLE. FOR EXAMPLE, A TRIANGLE WITH SIDE LENGTHS 2, 3, AND 4 IS OBTUSE. A TRIANGLE WITH SIDE LENGTHS 4, 5, AND 6 IS ACUTE.

We could try using the formula.

The triangle is five feet wide. But, how **tall** is it?

I don't think either of these slanty sides gives us its height.

To find its height, we need to measure from the top straight down.

95

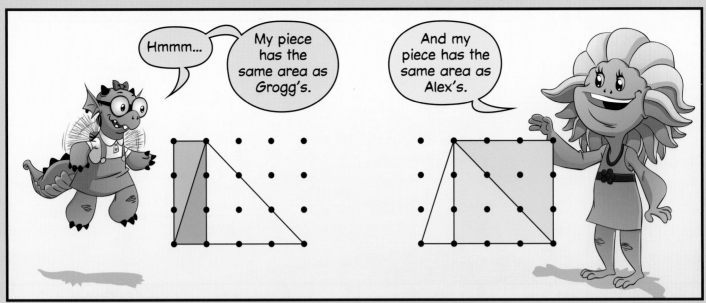

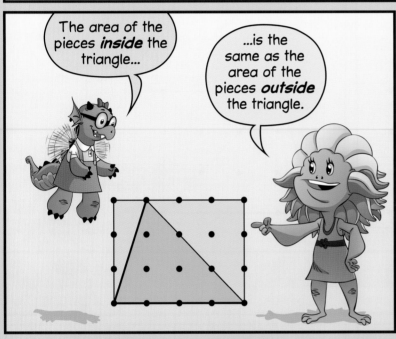

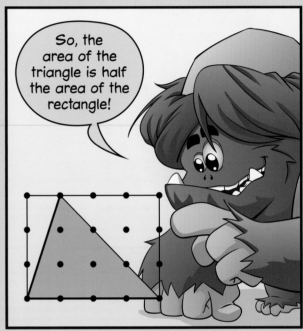

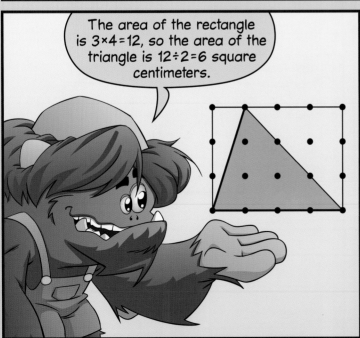

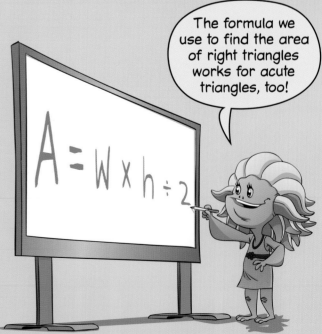

96

That's right!

You can find the area of any acute triangle by multiplying its width by its height and dividing by 2.

Does the same formula work for this obtuse triangle?

1 cm

1 cm

Does it?

Nope.

This triangle is definitely smaller than half the area of the rectangle around it.

The area of the rectangle is 5×2=10 square centimeters.

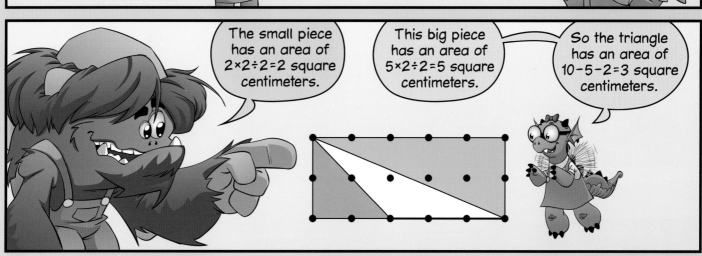

The small piece has an area of 2×2÷2=2 square centimeters.

This big piece has an area of 5×2÷2=5 square centimeters.

So the triangle has an area of 10−5−2=3 square centimeters.

Nice work!

The formula for the area of an obtuse triangle looks like this:

What's the base?

Area of an obtuse triangle
= length of base × height ÷ 2

$$A = b \times h \div 2$$

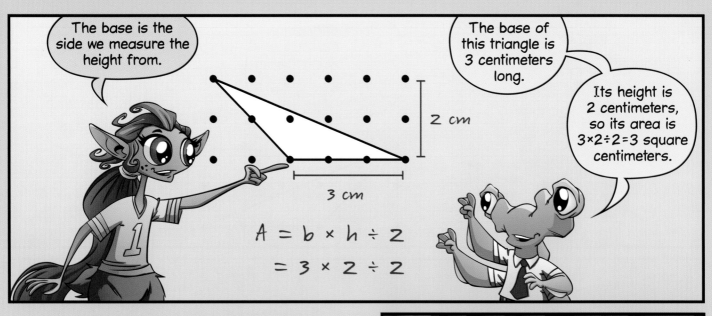

The base is the side we measure the height from.

The base of this triangle is 3 centimeters long.

Its height is 2 centimeters, so its area is 3×2÷2=3 square centimeters.

2 cm

3 cm

$A = b \times h \div 2$
$= 3 \times 2 \div 2$

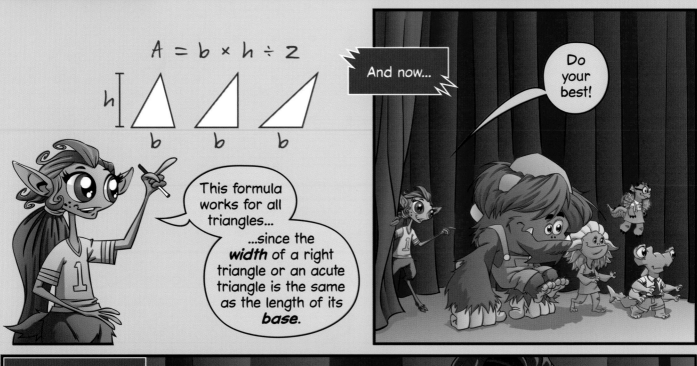

$A = b \times h \div 2$

h

b   b   b

This formula works for all triangles...

...since the **width** of a right triangle or an acute triangle is the same as the length of its **base**.

And now...

Do your best!

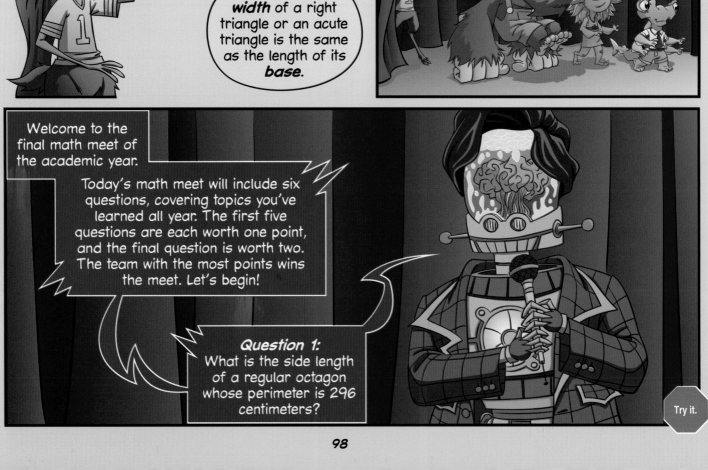

Welcome to the final math meet of the academic year.

Today's math meet will include six questions, covering topics you've learned all year. The first five questions are each worth one point, and the final question is worth two. The team with the most points wins the meet. Let's begin!

Question 1:
What is the side length of a regular octagon whose perimeter is 296 centimeters?

Try it.

98

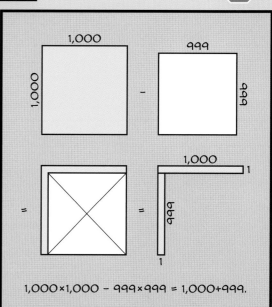

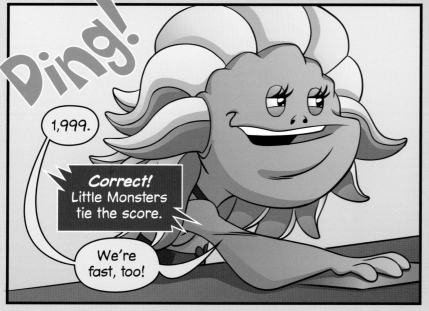

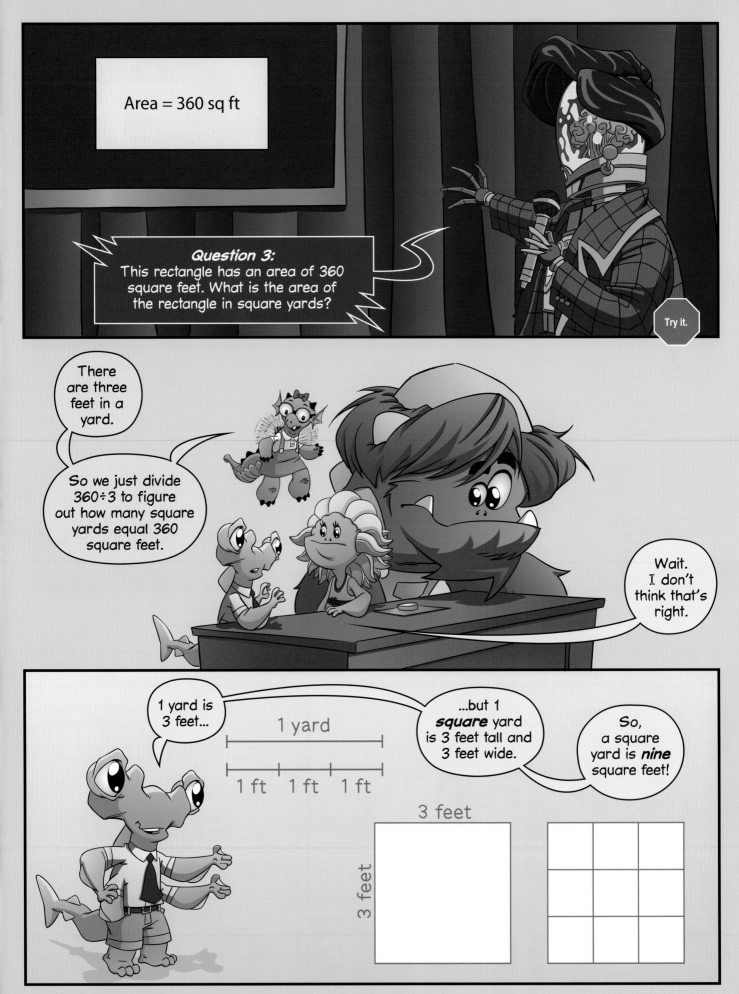

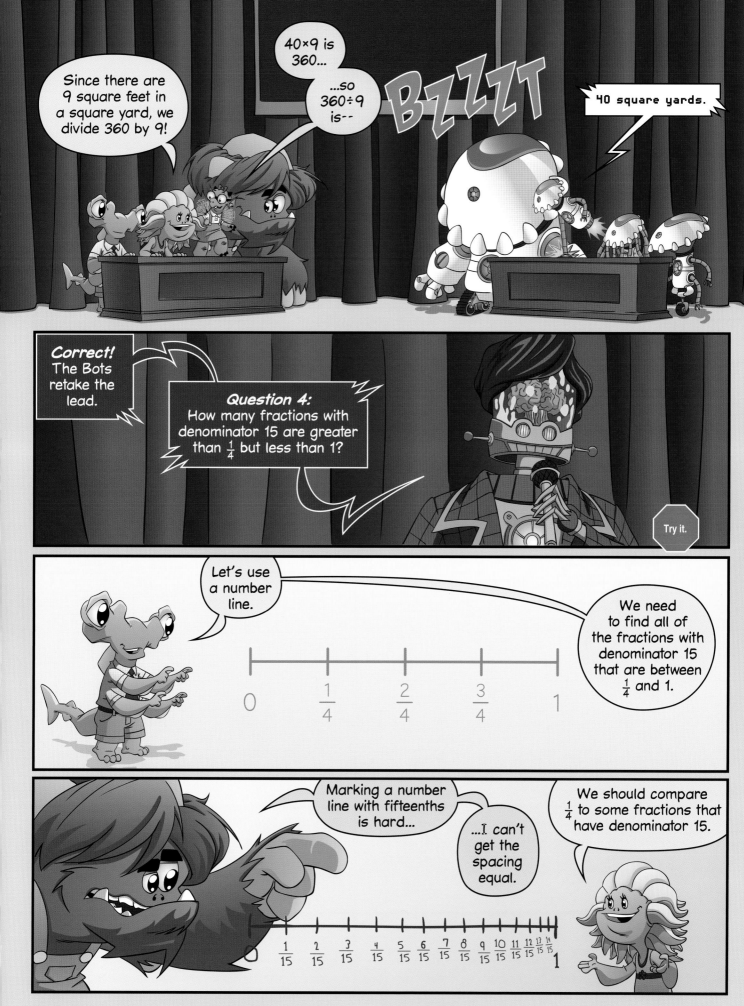

So, $\frac{1}{4}$ is between $\frac{3}{15}$ and $\frac{4}{15}$.

So, all the fractions from $\frac{4}{15}$ to $\frac{14}{15}$ are between $\frac{1}{4}$ and 1!

That makes a total of...

$$\frac{3}{15} < \frac{1}{4} < \frac{4}{15}$$

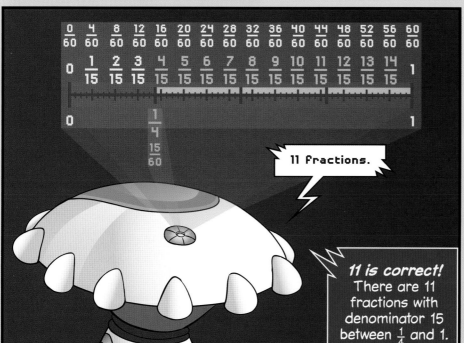

11 fractions.

11 is correct! There are 11 fractions with denominator 15 between $\frac{1}{4}$ and 1.

Whoa! That was smart. The Bots converted all of the fractions to sixtieths.

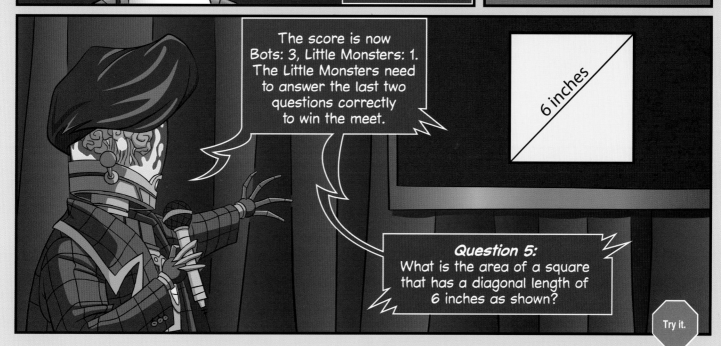

The score is now Bots: 3, Little Monsters: 1. The Little Monsters need to answer the last two questions correctly to win the meet.

6 inches

Question 5: What is the area of a square that has a diagonal length of 6 inches as shown?

Try it.

To find the area of a square, we need to know its side length.

But all we know is the length of its diagonal.

Maybe we can find the square's area by splitting it into shapes and rearranging them.

Like PolyGrogg!

If we cut the square along the diagonal...

...we get two right triangles.

We can put them together to make one big right triangle.

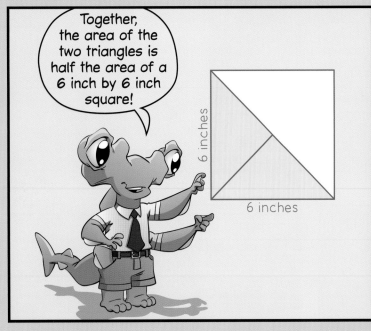

Together, the area of the two triangles is half the area of a 6 inch by 6 inch square!

6×6 is 36, and half of 36 is...

...18 square inches!

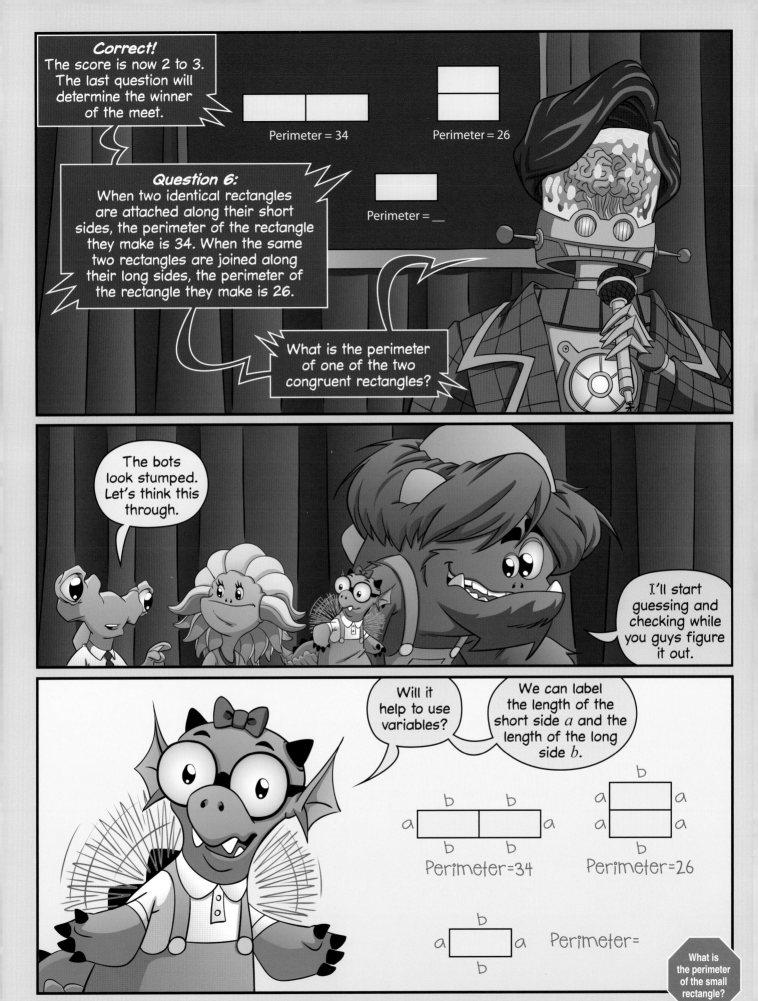

Correct!
The score is now 2 to 3. The last question will determine the winner of the meet.

Perimeter = 34

Perimeter = 26

Perimeter = __

Question 6:
When two identical rectangles are attached along their short sides, the perimeter of the rectangle they make is 34. When the same two rectangles are joined along their long sides, the perimeter of the rectangle they make is 26.

What is the perimeter of one of the two congruent rectangles?

The bots look stumped. Let's think this through.

I'll start guessing and checking while you guys figure it out.

Will it help to use variables?

We can label the length of the short side *a* and the length of the long side *b*.

Perimeter=34

Perimeter=26

Perimeter=

What is the perimeter of the small rectangle?

105

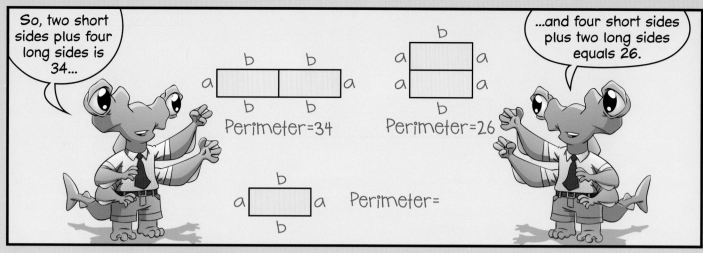

So, two short sides plus four long sides is 34...

Perimeter=34

Perimeter=26

...and four short sides plus two long sides equals 26.

Perimeter=

We can write these two equations.

$$a+a + b+b+b+b = 34$$
$$a+a+a+a + b+b = 26$$

If we **add** the two perimeters, we get **six** long sides and **six** short sides...

...with a total perimeter of 60!

$$a+a + b+b+b+b = 34$$
$$a+a+a+a + b+b = 26$$
$$6×a + 6×b = 60$$

If **six** short sides and **six** long sides add up to 60, then **one** short side and **one** long side add up to 10.

$$a+a + b+b+b+b = 34$$
$$a+a+a+a + b+b = 26$$
$$6×a + 6×b = 60$$
$$1×a + 1×b = 10$$

Since each small rectangle has **two** short sides and **two** long sides, its perimeter must be...

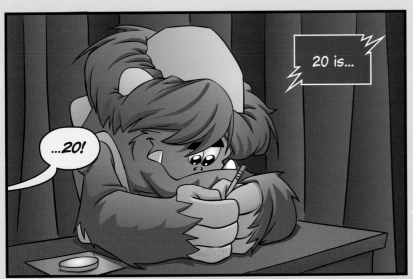

Practice: Pages 85-103

Index

For additional books,
printables, and more, visit

www.BeastAcademy.com